All Kinds of Weather

by Caroline Hutchinson

The weather is snowy. I like
to play outside where it is cold.
I can make a snowman.

What do you like to do?

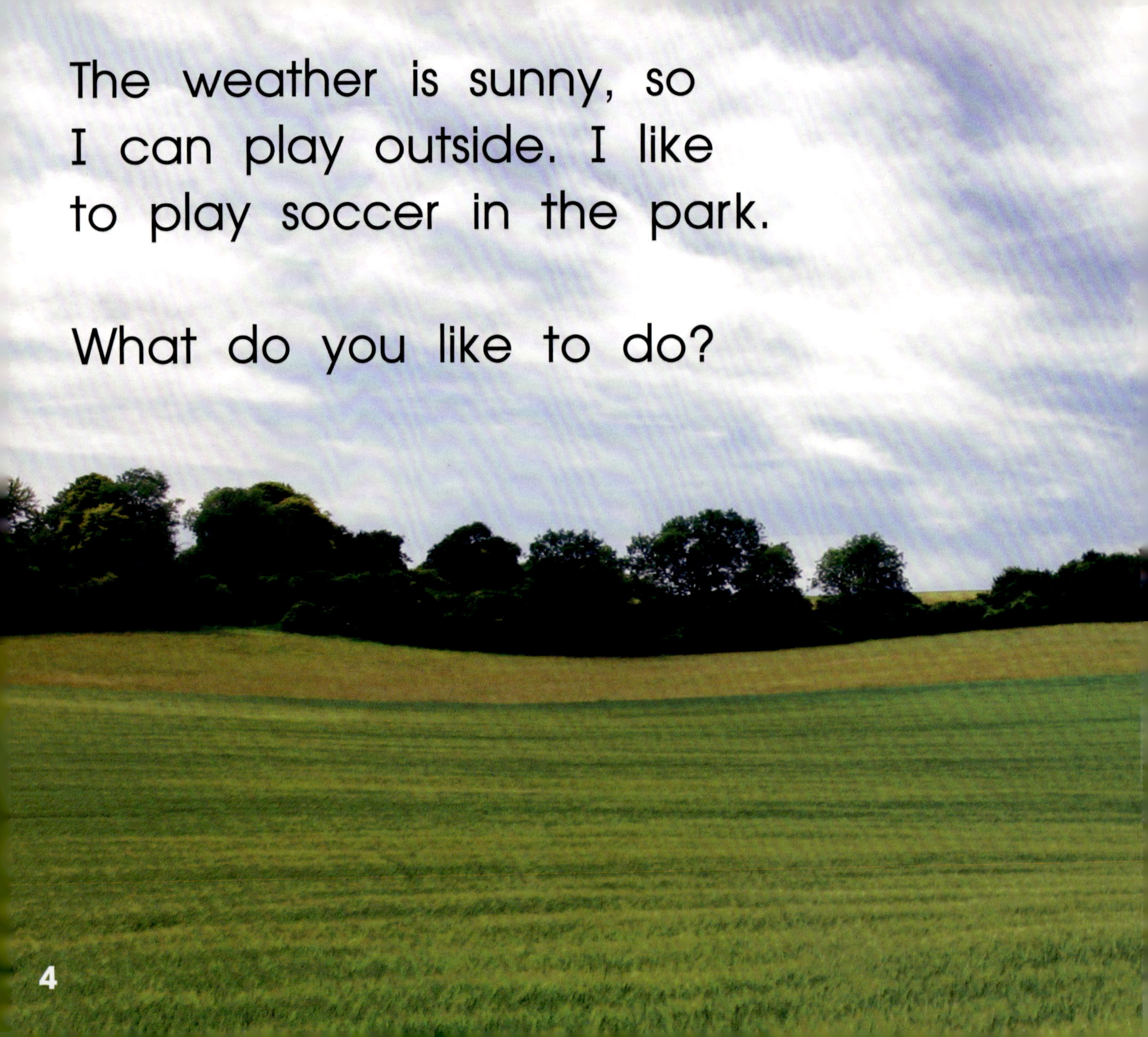

The weather is sunny, so
I can play outside. I like
to play soccer in the park.

What do you like to do?

1
9
2
19
15

The weather is rainy. I
play outside in the rain.
I can jump in the puddles.

What do you like to do?

The weather is windy. I like to play outside where I can fly my kite.

What do you like to do?

The weather is foggy. I stay inside and read a book.

What do you like to do?

The weather is stormy. I can see lightning. I stay inside because it is safe.

What do you like to do?

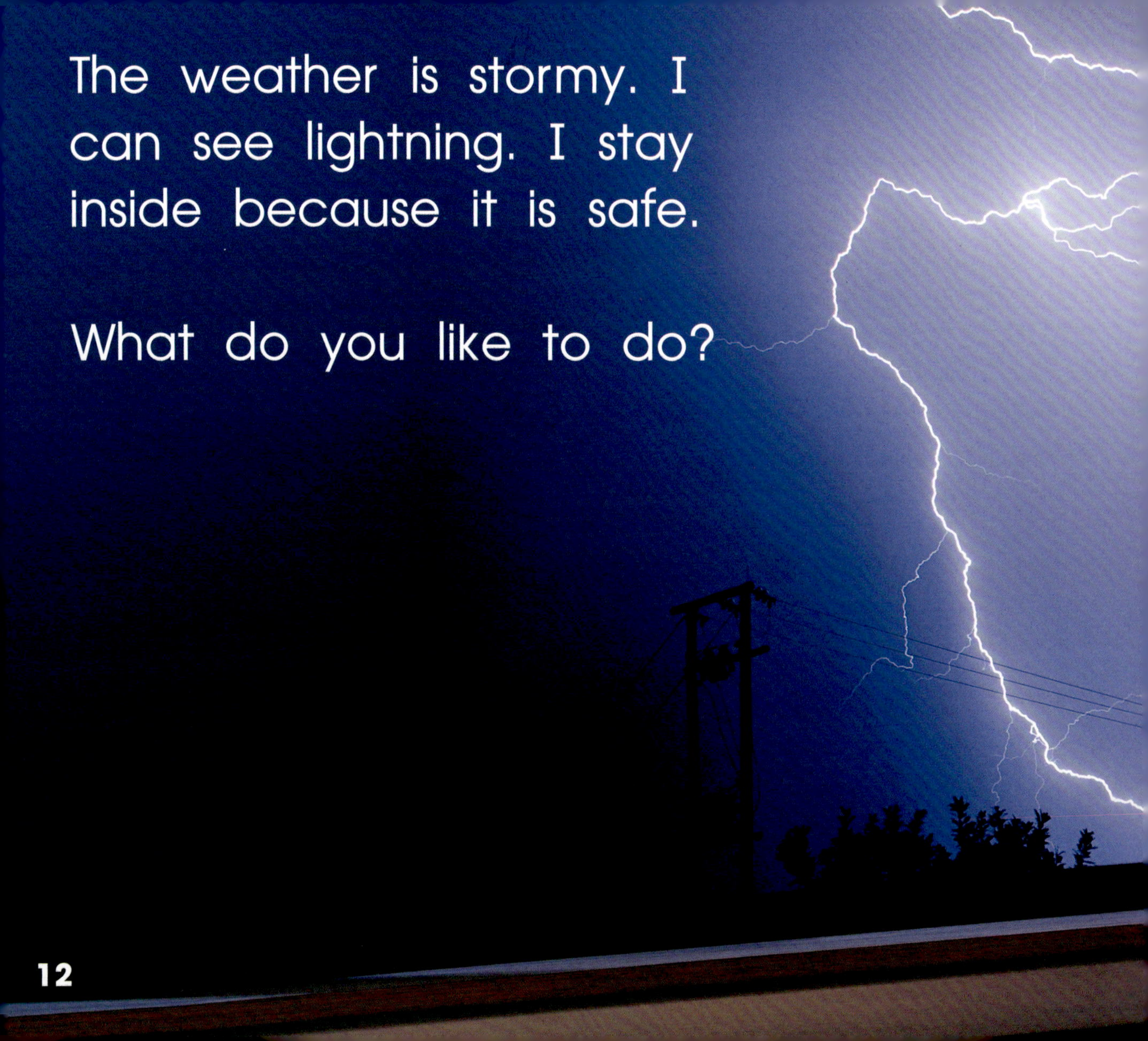

The storm is over, so
the sun is out now.
The sky is blue.

I can see a rainbow.

I can play outside!
What do you like to do?